This coloring book belongs to:

....

TEST COLOR PAGE

TEST YOUR COLOR SUPPLIES ON THESE ROCKETS TO SEE HOW THEY REACT TO THE PAGE

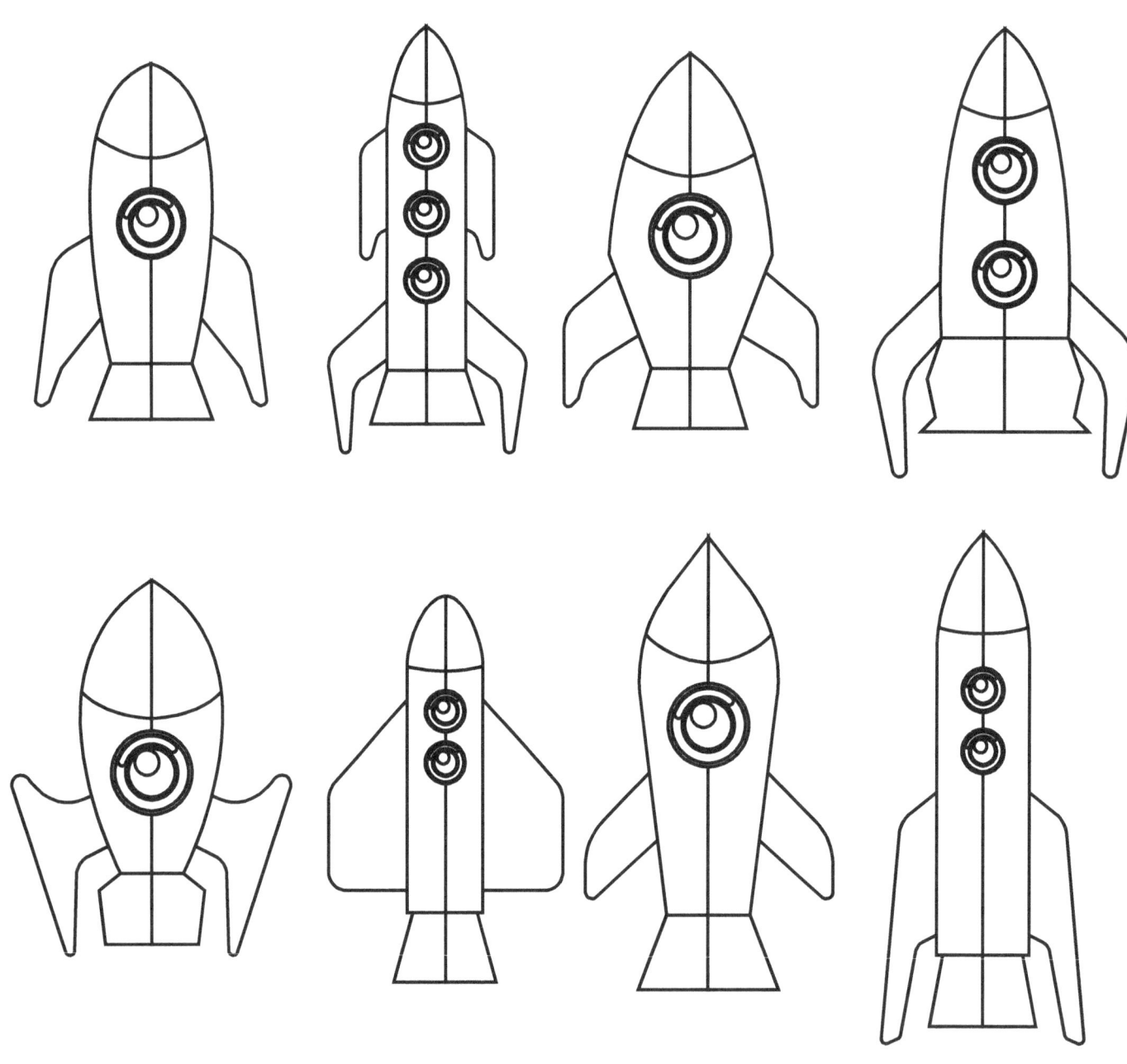

1. Space is completely silent.

Use Your Imagination

2. The hottest planet in our solar system is 450° C (842° F)

Use Your Imagination

3. There may be life on Mars.

Use Your Imagination

4. Nobody knows how many stars are in space.

Use Your Imagination

5. The sun's core temperature is 27 million degrees Fahrenheit.

Use Your Imagination

6. A full NASA space suit costs $12,000,000.

Use Your Imagination

7. Neutron stars can spin 600 times per second.

Use Your Imagination

8. There may be a planet made out of diamonds.

Use Your Imagination

9. The footprints on the Moon will be there for 100 million years.

Use Your Imagination

10. One day on Venus is longer than year.

Use Your Imagination

11. In 3.75 billion years the Milky Way and Andromeda galaxies will collide.

Use Your Imagination

12. If two pieces of the same type of metal touch in space they will permanently bond.

Use Your Imagination

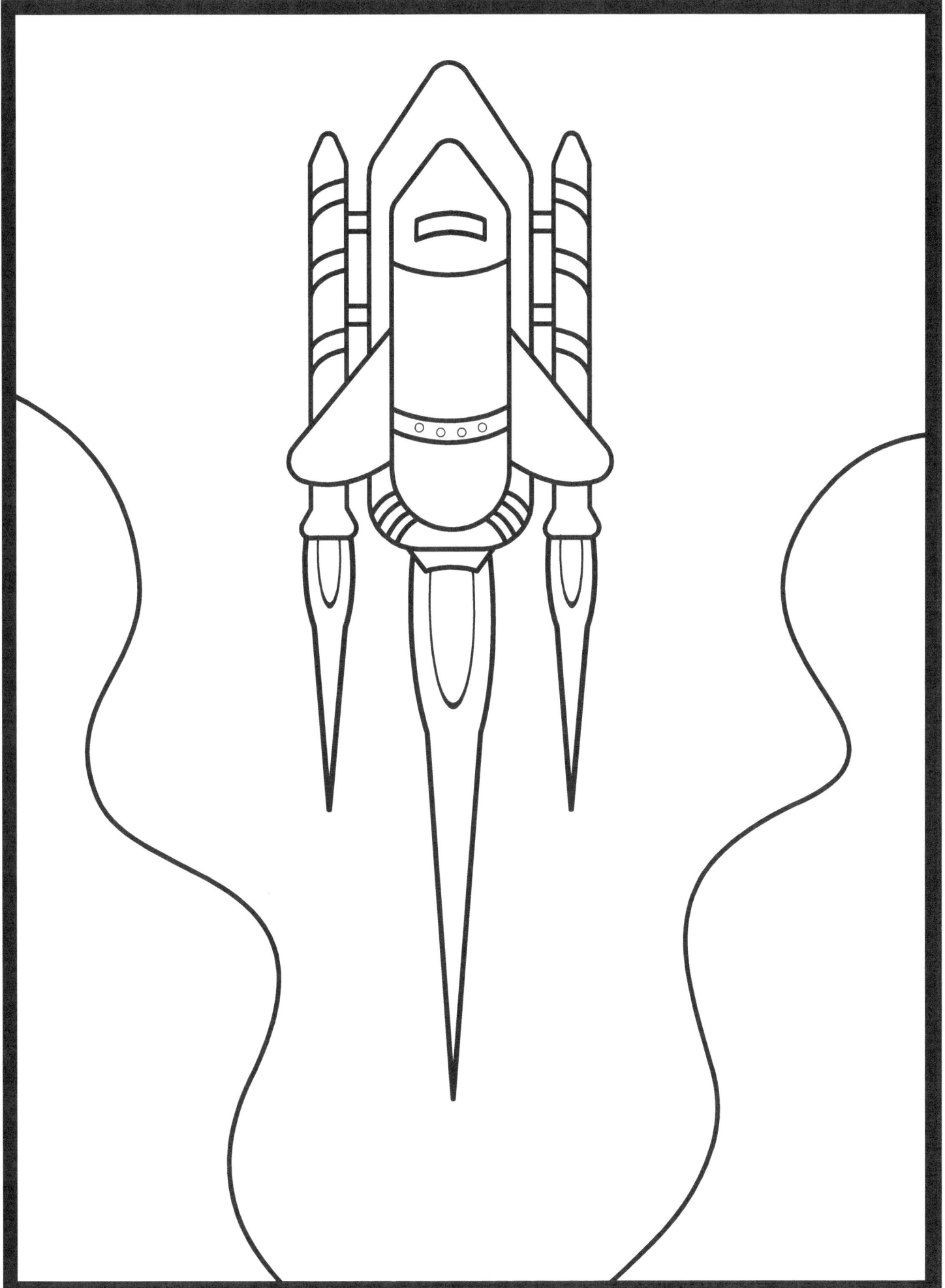

13. There is floating water in space.

Use Your Imagination

14. The largest known asteroid is 965 km (600 mi) wide.

Use Your Imagination

15. The Moon was once a piece of the Earth.

Use Your Imagination

16. The Sun's mass takes up 99.86% of the solar system.

Use Your Imagination

17. There is a volcano on Mars three times the size of Everest.

Use Your Imagination

18. Jupiter has 50 named moons, the most of any planet in our solar system.

Use Your Imagination

19. Meteorites are pieces of comets or asteroids that fall from space and land on Earth's surface.

Use Your Imagination

20 It takes 27 days for the moon to make one complete trip around the Earth.

Use Your Imagination

21. One million Earths could fit inside the sun – and the sun is considered an average-size star.

Use Your Imagination

22. You wouldn't be able to walk on Jupiter, Saturn, Uranus or Neptune because they have no solid surface.

Use Your Imagination

23. Many scientists believe that an asteroid impact caused the extinction of the dinosaurs around 65 million years ago.

Use Your Imagination

24. The Solar System formed around 4.6 billion years ago.

Use Your Imagination

25. Because of lower gravity, a person who weighs 200 pounds on earth would only weigh 76 pounds on the surface of Mars.

Use Your Imagination

26. The only planet that rotates on its side like a barrel is Uranus. The only planet that spins backwards relative to the others is Venus.

Use Your Imagination

27. Some of the fastest meteoroids can travel through the solar system at a speed of around 42 kilometres per second (26 miles per second).

Use Your Imagination

28. The first man made object sent into space was in 1957 when the Russian satellite named Sputnik was launched.

Use Your Imagination

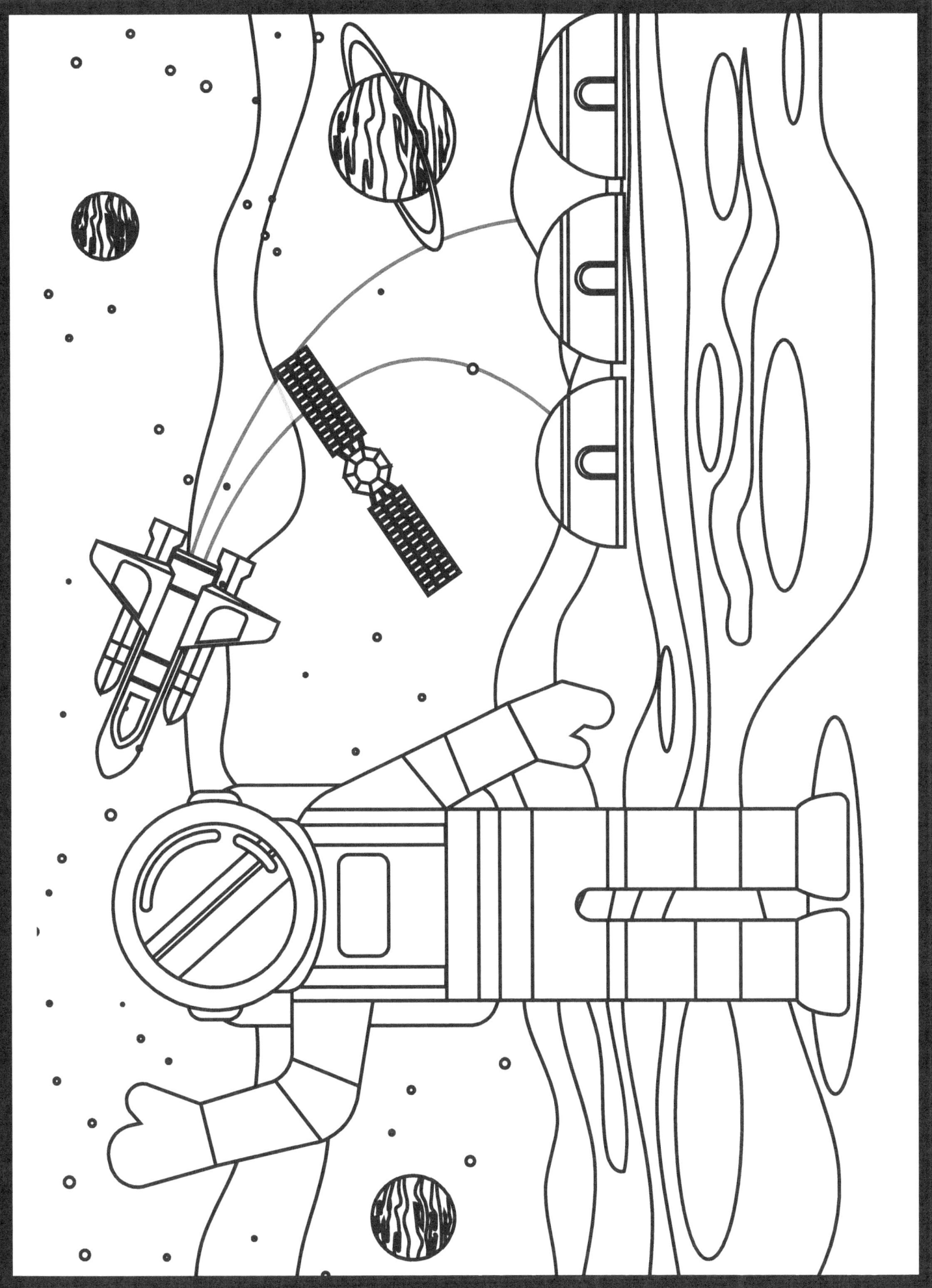

29. If you could fly a plane to Pluto, the trip would take more than 800 years

Use Your Imagination

30. There are more stars in the universe than grains of sand on all the beaches on Earth. That's at least a billion trillion

Use Your Imagination

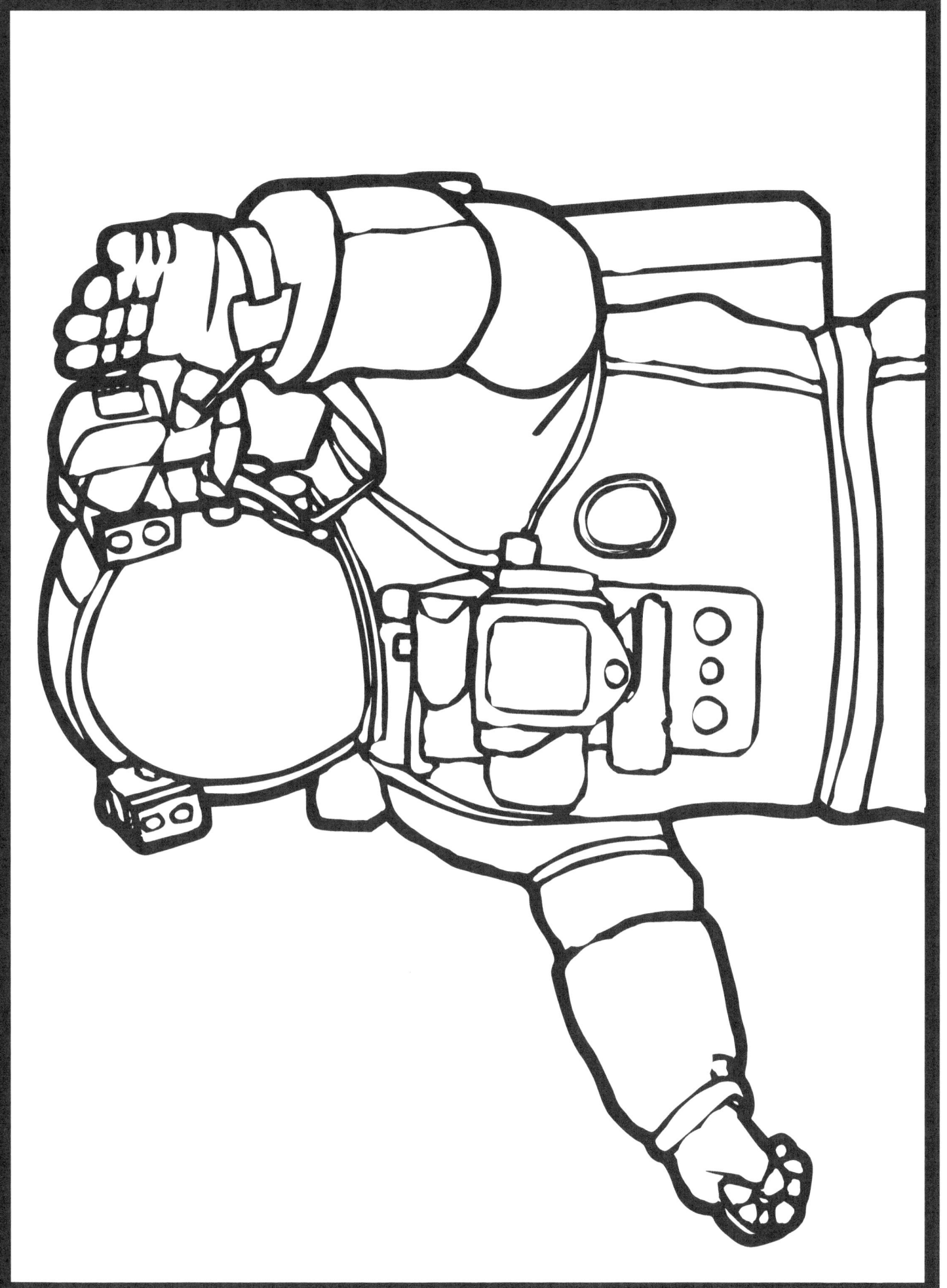

31. The sunset on Mars appears blue.

Use Your Imagination

What Did You Think of That Coloring Book?

First of all, thank you for purchasing this book. We know you could have picked any number of books , but you picked this one and for that we are extremely grateful.

We hope that it added at value and quality to your everyday life. If so, it would be really nice if you could share this book with your friends and family.

If you enjoyed this book and found some benefit in it, we'd like to hear from you and hope that you could take some time to post a review on Amazon. Your feedback and support will help us to greatly improve our designing for future projects and make this book even better.

We wish you all the best !!!